The Silent Science

A newly annotated version of

Mathematics in Life and Thought

by A. R. Forsyth

With a response essay by Kevin M. Kambo
Edited by Tristan Abbey

SCIENCE VENERABLE PRESS
TAMPA, FLORIDA
2024

Cover art: "Labor" by Raphael Sadeler I, after Maerten de Vos (1591),
Harris Brisbane Dick Fund, 1944 (Wikimedia Commons)

ISBN: 978-1-959976-05-9

Published by Science Venerable Press,
an imprint of Comarus Analytics LLC, Tampa, Florida

www.sciencevenerable.com

Interior designed by Jerry Janquart
Cover design by Nathaniel Roy
Art direction by Derek Thornton / Notch Design

"Knowledge cannot halt though her heroes fall."
— Andrew Russell Forsyth, 1897

Preface

Andrew Russell Forsyth (1858-1942) was never particularly famous and is barely known today. His mother passed away when he was only six years old and his marine engineer father died when he was a teenager. Nonetheless, riding a wave of scholarships and academic dedication, he made it to Cambridge—and dominated.[1]

The Silent Science is an homage, not a magnum opus—neither a biography nor a scholarly treatment. It merely contains at its core the text of a lecture, "Mathematics in Life and Thought," which Forsyth delivered in 1928 at University College of Swansea. This lecture has been edited only for formatting, chiefly breaking up the lengthy block paragraphs that were in style at the time, and is accompanied here by new annotations and appendices that dig a bit deeper than is possible in a footnote. I am also delighted to feature a response essay by Kevin Kambo, professor of philosophy at the University of Dallas.

This project would not have been possible without the research assistance provided by Barrett Hellmann. Further thanks are due to Benjamin Brady, Stephen McKeown, and Mark McKinzie. The MacTutor History of Mathematics Archive at the University of St Andrews remains a wonderful resource. HathiTrust Digital Library, Perseus Digital Library, Wikisource, and the Internet Archive are also invaluable aids.

—Tristan Abbey
Tampa, Florida

---

1 See E. T. Whittaker, "Andrew Russell Forsyth, 1858–1942," *Biographical Memoirs*, Vol. 4, Issue 11 (Royal Society, November 1942), pp. 208-227.

Andrew Russell Forsyth
(Credit: Royal Society)

Table of Contents

Mathematics in Life and Thought

Lecture Delivered at University College of Swansea[1]
November 14, 1928

A. R. Forsyth
FRS, FRSE[2]

1 University College of Swansea operates today as University of Wales Swansea.
2 Fellow of the Royal Society, Fellowship of the Royal Society of Edinburgh.

I

There is an ancient adage, certainly as old as the naturalist Pliny [the Elder], who perished in the famous eruption of Vesuvius [in A.D. 79], that "a shoemaker should not look above his last."[1] The wisdom of the adage is always being ignored. The newspapers give evidence of the neglect in their daily spate of ingenuous letters. Every publishing season adds to the evidence by the recurring deluge of new books which seem no nearer an end now than they were in the days of the troubled Preacher who, like many another student, found much study a weariness of the flesh.[2]

The neglect is not unknown even in the range of occasional lecturers of the strictest academic credentials. Physicists, whose main concern has been with the inanimate matter of the universe, have been heard to pontificate on the origins of life and its continuity after death. Men of science will discourse on philosophy with no more success and no more clarity than are achieved by Beatrice in Dante's *Divina Commedia.*[3] Philosophers, in all the

1 Pliny the Elder describes how a certain Greek painter named Apelles would hide behind his paintings to overhear constructive feedback from passersby: "It was under these circumstances, they say, that he was censured by a shoemaker for having represented the shoes with one shoe-string too little. The next day, the shoemaker, quite proud at seeing the former error corrected, thanks to his advice, began to criticize the leg; upon which Apelles, full of indignation, popped his head out, and reminded him that a shoemaker should give no opinion beyond the shoes."

2 "Much study is a weariness..." See Ecclesiastes 12:10-13.

3 In Canto XXIX of *Paradiso*, Beatrice herself chastises philosophers and

precision of metaphysics, will enunciate canons for the progress of observational science. Nay, theologians have been known to formulate fundamental principles which, as being the eternal verities, are the unchanging limits of scientific attainment. Even the citizen in the bus will settle the policy of a nation, though his knowledge of an issue at stake could be written on a picture postcard, leaving plenty of space for the picture.

So let me, very respectfully, for once be the obedient shoemaker of the adage, not look above my last, and offer you a sample of my own wares.

The subject of mathematics, to some aspects of which my lecture will invite your attention, is regarded variously: sometimes with respect, rising through awe to wonder; sometimes with dislike, degenerating through repulsion to hatred. Occasionally it is regarded as a key, even as the guide, to trustworthy knowledge; sometimes as the dismal process of calculations to be performed by hired computers.[4] In any event, why take it as the matter of a lecture?

For there are at least two diverging views as to the nature of such a performance. One view is that the lecture should give a tabloid summary of what is hidden in books that are not read: yet, in the case of my choice, who wants a summary of what

---

preachers of the world for behaving just like Pliny the Elder's shoemaker:

> On earth men dream, therefore, while they're awake
> Some in good faith, and some deceitfully;
> Of guilt and shame the greater share these take.
>
> Ye on the earth, in your philosophy,
> Are not for long content to tread one path,
> Enamoured of vain show and subtlety...
>
> All men, to show their ingenuity,
> Contrive their own inventions—these they preach;
> The Gospel is passed over silently.

4 Computers: individuals paid to perform calculations.

seems the infinitely arid and the absolutely incomprehensible, as so obviously appears whenever a mathematical book is opened?

Another view is that the lecture should give what is not to be found in any of the books, so that the exposition of novelties shall induce, in a sympathetic listener, the frame of mind that is willing to acquire fuller knowledge through patient labour. On this view the listener must indeed be sympathetic, very willing to provide the patient labour for the acquisition of mathematical knowledge: for he will find, as Euclid told a bored and discontented pupil in words that have lived for more than two thousand years, "There is no royal road to Learning."[5]

My purpose differs from both of these implied aims. It is to invite you to note one or two of the uses, some simple, some recondite, of mathematics, and, by noting them, to attempt an appreciation of the issues which are forced into daily discussion and daily action.

For the ordinary course of affairs a great English divine declared that "probability is the very guide of life."[6] Usually certainty is lacking: and probability can be taken as the only reasonable guide amid an occasional maze of possibilities. Yet, when sifted and analysed, this probability often reduces itself to a more or less mathematical (chiefly, I fear, less mathematical) measure of moral and material advantages to follow. The spirit of mathematics, if not its method, dominates the sifting analysis: we can sometimes regret that the method is lacking.

Let me offer a couple of illustrations.

Some of you may remember a legend of wide circulation in 1914 before the Great War was many weeks old: how a force

---

5 Proclus: "It is also reported that Ptolemy [I of 3rd-century B.C. Egypt] once asked Euclid if there was not a shorter road to geometry than through the *Elements*, and Euclid replied that there was no royal road to geometry."

6 Joseph Butler, *The Analogy of Religion* (1736): "For nothing which is the possible object of knowledge, whether past, present, or future, can be probable to an infinite Intelligence; since it cannot but be discerned absolutely as it is in itself—certainly true, or certainly false. But to us, probability is the very guide of life."

of 80,000 Russians had been landed in the north of England, transported from the northeast coast to Portsmouth, and thence shipped to France.[7] The moral support of those Russians was a comforting belief at the moment. There had been days, only a human generation earlier, when everything Russian had been anathema to patriotic Englishmen: but those days were past. Though we might be martially arrogant enough to feel that we did not need the help, it still was good to think that we had it: so comforting that the news was accepted without question. Retired military men in clubs, an unfailing source of trustworthy wisdom, were supposed to know it: the unverified gossip of the smoking-room gathered confidence in statement as it spread: the good news was believed. I was staying on the Dorsetshire coast at the time: and during a week-end a visitor brought information from a London club that the station-master at Dorchester had seen the Russians pass through his station in the train.

My mathematical scepticism was roused: my informant, by successive queries mildly posed, was induced to estimate the number of men in a compartment, the number of compartments in a carriage, the number of carriages in a train even in a period of emergency, and the consequent number of trains needed to convey the said 80,000 Russians. When, on the basis of his own estimates, he informed me (correctly) that over one hundred trains would be required for the transport, I asked him whether the Dorchester station-master had seen over one hundred trains pass through his station. His answer was blank silence: for him, as for me, the Russians vanished into thin air—a little salutary arithmetic had dissipated a hopelessly false legend that was fostering false confidence. Probability had not been the guide of my informant's judgment.

7 The First World War began on July 28, 1914, with Austria-Hungary's declaration of war against Serbia. "A Russian army of 72,000 men, transported from Archangel, the most northern part of Russia, was landed at Aberdeen, on the east coast of Scotland…." See *New York Times*, September 4, 1914.

I suppose that, in Swansea as in other communities, considerable industry is devoted by eager optimists to competitions anticipating the results of football matches. Newspapers (all of them prudent enough to insure against loss in the venture) vie with one another in offering large sums of money for merely filling up slips of paper: in those slips entries are to be made estimating the result for each one of a dozen matches, a win, or a draw, or a loss, for a specified side. To the person who anticipates all the twelve results correctly, a large sum of money is paid. So the competitors, eager for wealth but niggardly in constructive labour, spend long periods in hatching their prophetic guesses. It is not a game: it is not a business; it is a gamble.

I sometimes wonder whether the competitors have the faintest idea of the minuteness of the chance of individual success in the competition, if it can be dignified by such a title. There are more than four thousand different ways of getting every guess wrong; there are more ways of getting eight guesses wrong and the other four right than there are of getting any other distribution; there are more than half a million ways of having one or more errors in the selection: there is only one way of getting them all right.[8] Need I say more? To this fatuous occupation, Bishop [Joseph] Butler might have hesitated before declaring probability to be the very guide of life.

8 See Appendix: Gambling Probabilities.

II

Let us pass to consider matters somewhat more serious and more precise.

A copious definition of mathematics can be given by describing it as the craft of counting as in the arithmetic of daily life, as the craft of measuring as in the original needs of surveying and engineering, as the craft of observing and noting as in the early days of astronomy. When the simple processes of counting, of measuring, and of recording observations are pursued systematically, it is found that the repetition of some particular issue requires the particular repetition of some unchanging and appropriate process—a rule is evolved: the rules aggregate into a science. Moreover, there arises the necessity for some discrimination between the nebulous inspiration of a guess, which in our graver moods we call an "idea," and the comparative certainty of a systematic inference, which those graver moods dignify with the title of a "theory." As a mere fact, speculations and ideas may be almost as old as the hills: laborious methods can lead to a theory which, new to-day, may be discarded or be confirmed tomorrow in the light of fuller knowledge.[9]

Again let me offer a couple of illustrations, more academic than imaginary Russians or a gambling quest for an unearned increment. One shall be from astronomy, the other from the structure of matter or (as the old Latin poet Lucretius styled it) the nature of things.[10]

9 "As old as the hills." See Job 15:7.

10 *De Rerum Natura* (On the Nature of Things):

Systematic astronomy had its establishment in observations as accurate as time and patience and simple instruments would allow. A marvellous collection of facts, relating to such abstruse phenomena as the sequence of eclipses of the sun and of the moon, was made by the ancient Chaldeans.[11] Later the Greeks, whose passion for geometry was almost a rule of life, began to describe the heavens geometrically as they saw and we see the heavens. The natural, the almost inevitable, consequence was that, because the Greek observers described accurately what they saw, they made the earth the centre of the universe. (Here let me say that these old Greek descriptions remain as accurate as man could make them: where we part company from the Greeks is in the assumption, obvious but untrue, that the phenomena require the earth to be the centre of all things.)

Yet nearly three centuries before the Christian era the Greek philosopher, Aristarchus of Samos, was an early astronomical heretic—he groped towards a heliocentric theory of the universe: with him it was a speculation, an idea, based upon an inference that the sun is seven times as large as the earth, at a time when there was little knowledge of the size of the earth and less of the size of the sun.[12] His speculation remained in the

---

So true it is no solid form is found.
But yet because true reason and nature of things
Constrain us, come, whilst in few verses now
I disentangle how there still exist
Bodies of solid, everlasting frame—
The seeds of things, the primal germs we teach,
Whence all creation around us came to be.

11 A Mesopotamian people typically associated with Babylonian astrology, similar in concept to the "magi from the east," "wise men," etc.

12 Archimedes notes of Aristarchus: "His hypotheses are that the fixed stars and the sun remain unmoved, that the earth revolves about the sun on the circumference of a circle, the sun lying in the middle of the orbit, and that the sphere of the fixed stars, situated about the same centre as the sun, is so great that the circle in which he supposes the earth to revolve bears such a proportion to the distance of the fixed stars as the centre of the sphere bears

region of fancy: the traditional trust in Aristotle maintained the acceptance of the old geocentric theory; there were not enough facts to achieve the tragedy of killing a theory.[13] In their turn the theologians of the West, conservative as are all orders of organized priesthood, made the geocentric theory an article of orthodox faith: and for centuries the universe, thus theocratic in matters outside religion and faith, could be described only in the vocabulary of the schoolmen in theology. A Catholic monk, [Nicolaus] Copernicus, opened the path to a change of estimate, though the innocent dedication of his book to the Pope [Paul III] saved his name from contumely: had he lived, it might not have saved him from the fate of [Galileo] Galilei. Gradually, almost painfully, methods systematized from facts killed the old theory: and the inspired guess of Aristarchus passed from the range of happy surmise into a knowledge of the solar system of which even the most exuberant imagination had not dared to dream.

My second illustration is taken from a more speculative range of thought—more speculative because less openly amenable to observation and to experiment. Many of you doubtless are familiar with the recent predominance of the word "atom" in scientific discussions. There was a time, even now easily recalled, when the use of the word was an implicit declaration that finality had been attained: human knowledge could not penetrate the indivisible. In the interval, a school of mathematical physicists has devoted itself to the study of the atom: its apostles have concluded that, mathematically, an atom is a universe in itself, a sort of solar system with recurrent revolutions of its members that can be subjected to calculation.[14] Different atoms, in present speculation,

---

to its surface." For the comparative sizes of the Sun and Earth, see Appendix: Diameters of the Sun and Earth.

13 *On the Heavens*: "Because there must be something at rest at the centre of the revolving body [i.e., the heavens]...Earth then has to exist; for it is earth which is at rest at the centre."

14 In this classic view, the atom consists of a nucleus, representing the Sun, and electrons, which are planets orbiting around it.

are different universes which, under the compulsion of electric bombardment, may lose some of their constituents and, in doing so, become a changed universe.

But mark the history. The atom was the irresoluble element by which Democritus, four centuries before the Christian era, explained the phenomena of matter to one of the most critical nations of thinkers the world has known: the atom became the explanation of animate, as well as of inanimate, nature. The culmination of that ancient theory came in the doctrine of Epicurus; and its fine exposition by Lucretius is nearly two thousand years old. In varying forms, with changing adaptations, it was tried, was found wanting, was driven from pillar to post, until it seemed to have reached a secure haven in chemistry.

Where is the atomic theory now? Thirty years ago chemists were shocked by the resolution of the atom, the absolutely irresoluble element; now the more elusive electron has taken the field. The scientific world is in pursuit of the electron: Sir Ernest Rutherford is battering the atom by an electric attack under staggering voltage; and the negative electrons thus violently expelled from the domain of the positive nucleus are realizing the speculative guess made, centuries ago, by Hipparchus, that atoms could lose portions of themselves and thereby change their essence.[15] Sir Joseph Thomson, the brilliant master of Rutherford, goes further than his pupil and suggests that the electron itself is a universe, perhaps another mathematical solar system; and only two short months ago the distinguished son of Sir Joseph Thomson was able, as the result of investigations, to compare the size of the electron and that of the atom, neither of which has been seen by man.[16]

---

15 Reference to the practice of "bombarding" the atom with high-energy "alpha" (α-) particles, each approximately 5 million electron volts (MeV). The reference to Hipparchus, who left no trace of his views on chemistry, is probably erroneous. A better candidate is Hippocrates (via Galen): "Surely something that is to feel pain must necessarily meet these two requirements: it must be capable of change and capable of sensation."

16 Rutherford studied under Thomson at the Cavendish Laboratory from 1895

Physical speculation is outrunning its mathematics: but, as so often before, the return to mathematics is ever made as soon as there are facts amenable to calculation. The facts remain, and the calculations based on the facts remain. Theories spring up and flourish while they conform to the facts and the calculations: they wilt, when new facts come that demand requirements which the theories cannot supply. It resembles the historic cult in the Arician grove: an old theory of yesterday is forsaken for a new one of to-day, itself to be forsaken to-morrow, like

> The priest who slew the slayer
> And shall himself be slain.[17]

---

to 1898. Thomson originally called the electrically charged particles that he discovered "corpuscles," but "electron" quickly became the convention: "[E]ach corpuscle may be said to extend throughout the whole universe, a result which is interesting in connection with the dogma that two bodies cannot occupy the same space." The son referenced here is George Paget Thomson.

17 See Appendix: Cult of Diana.

III

What part does the science of mathematics play in all this scientific tornado? And, indeed, what does the science include, and what may it claim to achieve?

Broadly speaking, as soon as any progressive subject attains a stage where its phenomena admit of some kind of measurement, then and there the science of calculation can begin to deal with the measurements: thereafter the subject can be assisted by mathematics. Calculations can predict some results that must ensue from combinations of facts, and so can render some experiments superfluous. Calculations can even anticipate experiment and observation: such has been the case with [William Rowan] Hamilton's conical refraction in optics[18] and with the electric waves which first emerged from the mathematical labours of [James Clerk] Maxwell and [Heinrich] Hertz. Calculations can utilize facts in non-experimental fashion and produce results that lay concealed. But no calculation can produce results that do not belong to the range of the subject-matter: thus astrology, and alchemy, and necromancy, bound to be scientific failures, were and are successful frauds owing to the undying gullibility of human nature in its wistful yearning for the unknown and the unattainable.[19]

---

18 In 1832, Hamilton predicted that a certain type of crystal could refract a single ray of light into a cone of rays. This prediction was confirmed later in the same year.

19 Possible allusion to Harry Peck in *Cosmopolitan* (1899): "[C]ollective man to-day…is cursed with a vague disquietude, a discontent with everything

Briefly, the science of mathematics cannot be a substitute for essential experiment: but it can show how experiments and observations, duly systematized, can be elucidated so as to discriminate between what is principle and what is detailed consequence of principle. Sometimes it can lead, though its guidance is unrecognized, to simple devices which economize labour to an extent almost beyond credibility. Thus some of you know the slide-rule by practice; many of you will know it by name; but I am not sure that every student, or even every professor, would be prepared to furnish a reasoned statement justifying the various processes for which that admirable instrument is used. Yet it can be regarded as a compendium of logarithms, without tables and without any necessity for calculation: and logarithms are but a means of turning abstruse multiplication into simple addition and abstruse division into simple subtraction.

Again, your Russian peasant usually is illiterate: but he will make out his invoice for wool, or furs, or timber, or corn and rye, as accurately as your western clerk by surprisingly simple means that mark a limit of his attainments. He can multiply by the number two, but by no other number; he can divide by the number two, but by no other number; and he can perform simple addition.[20] By these three operations he effects all that he needs for business transactions, and he effects it accurately. But he could not give you one word of explanation of a process which to him has become a habit; and if you explained the mathematical argument that justifies his action, he could not understand one word of the explanation. Does his process strike you as wonderful? It is no more wonderful, it is less wonderful, than the process employed by the smallest shopkeeper. Your Welsh shopkeeper calculates in what mathematicians call the scale of four for

that is, a yearning for the unknown and the unattainable."

20 The mathematics literature of this period is replete with references to the so-called Russian Peasant Method of Multiplication, both as a tool for instruction and as the subject of proofs to explain why the method works. See Appendix: The Russian Peasant Method.

farthings, in the scale of twelve for pence, in the scale of twenty for shillings; he adds and he multiplies in the scale often. But he knows nothing of numerical scales; and probably he would not understand a word of reasoned explanation of such scales.

The Russian peasant actually calculates, in essence though not in obvious form, in the scale of two: and some kind though unknown genius, in a dim and distant past, has turned the simple mathematics of the scale of two into the Russian peasant's process of arithmetic.

IV

And if such is the fact in simple matters, you need not be surprised if the science of mathematics is used to high economic purpose where, at some stage, its calculations are obtrusively significant. Another couple of examples will justify such a statement.

At the present time, parts of the country are soon to be covered with a vast network of wires, stretching from pier to pier, for the transmission of electric power from central stations to relatively remote places where the power cannot be generated and cannot otherwise be obtained or concentrated. All sorts of complicated questions of an engineering kind arise, as fundamental in their way as questions connected with the construction of a liner to cross the Atlantic or a bridge to span the Ganges: questions of stress, of strain, of strength to face unknown but imaginable conditions which may become grimly real; and the answers are the results of mathematical calculations.[21] These calculations are

21 In 1927, the fastest ocean liner was RMS *Mauretania*, which could make 25 knots, and the largest was RMS *Majestic*, with a gross tonnage of some 56,000. The legacies of both ships intertwine with World War I: the sister ship of *Mauretania* was the RMS *Lusitania*, sunk by Germany in 1915, and the *Majestic* was originally a German liner, given to the British in reparations for the wartime sinking of a hospital ship. In comparison, the gross tonnage of modern Carnival Cruise Line ships range from 70,000-180,000; speed has improved only marginally to a maximum of some 30 knots. The Hardinge Bridge, which crosses the Lower Ganges in present-day Bangladesh, was opened in 1915 to great acclaim in engineering circles. Made of steel and stretching over one mile, it was among the longest in the world at the time. See Robert Richard Gales, "The Hardinge Bridge over the Lower Ganges at Sara," *Minutes of the Proceedings of the Institution of Civil Engineers*, Vol.

made in an office, but they do not remain in the office: they are utilized to indicate critical tests and to adapt given possibilities for the making of those tests in a variety of external conditions. Not until all the tests are satisfied would the engineers be justified in proceeding to their work: and engineers know, and can imagine, better than all others concerned, how much superfluous labour has been saved by the mathematical calculations.

There were days when the telegraph, overhead or submarine, was at once a new toy and a new mystery.[22] Once the springs of action were known, their doings were regularized by instruments devised on a mathematical basis. Systematic processes were thus possible: their mathematical basis was worked out by Lord Kelvin (then Professor William Thomson) especially for submarine cables; and through his double genius for mathematics and for engineering applications of mathematics, the telegraph took its place in the routine of daily life.[23] Now it is so ordinary, almost so stale and so old-fashioned, as to seem on the verge of supersession. For you are surrounded by wireless telegraphy at every turn; the beginnings of telephotography are being made whereby you may see in your evening paper a transmitted photograph of some event occurring only a few hours earlier perhaps some thousands of miles away.[24] Does it ever occur to jaded satisfaction to inquire how so much has come into ordinary life which even in our grandfathers' age would have been declared impossible or fanciful? The story is not long, though the labour has been multitudinous.

Probably any historian of modern electrical science would be prepared to acknowledge [Michael] Faraday's researches as the beginning of a new epoch. Great as was his genius, and valuable

---

CCV (1920), pp. 18-67.

22 The first long-distance telegraphs: overhead from Baltimore to Washington, D.C. (1844) and underwater across the English Channel (1851).

23 Lord Kelvin developed the fundamental "telegraph equation" and engineered devices necessary to its application. He was not related to J. J. Thomson of electron fame.

24 The Associated Press launched its "wirephoto" service in 1935.

as were those researches, Faraday left them largely as a treasury of rich results, fundamental in bearing, capable of ordered development from their experimental stage and wisely formulated laws, but still undeveloped. Then another genius with another bent, Maxwell the mathematician, regarded the quantitative results obtained by Faraday as the material for systematic mathematical analysis; and the monument of his life is his ordered mathematical exposition of a new mathematical theory of electricity and magnetism founded on the Faraday results. But, as so often has been the fact in the movements of human knowledge, when once the accurate scientific observations received their systematic treatment by mathematics, interesting possibilities were revealed that hitherto had not even been surmised; and one immediate consequence of Maxwell's work was his creation of an electro-magnetic theory of light, through the identification of a relation between certain electrical constants and the speed of the transmission of light.

The notion of waves and vibrations now entered boldly into the discussion. Another mathematical genius, Hertz, developed the consequences of the Maxwell theory and, by his calculations, obtained inferences in advance of observed results. The constructive genius of engineers—[Guglielmo] Marconi is a typical and outstanding example—was fired to new efforts in practical application. Soon there was the initial sending of messages through space (call it the ether, if you choose, but do not suppose that a name implies knowledge) from one land to another, from land to ocean and ocean to land: last Sunday's armistice service in Trafalgar Square in London was heard in New Zealand.[25]

Let me leave you in the middle of sounds, which may come from a studio near at hand or a continental orchestra perhaps in

25 Ether was once believed to be the substance through which light propagated. By this time, the accumulated weight of some four decades of theory and experiment challenged its existence as a material concept. The Armistice event broadcast around the world occurred on November 11, 1927—Friday, not Sunday—and was held at Albert Hall and nearby Hyde Park, not Trafalgar Square.

Rome or Madrid: and in all your surprise and happiness do not entirely forget that, throughout the development, the science of mathematics has played, is playing, and will continue to play, its significant part in every endeavour.[26] Electrical engineering is not wholly electrical, in practice or in theory: its successes are achieved not solely through its machinery, however elaborate. Behind all its progress, as part of the reasoning mind of it all, there labours the science of mathematics as an angel of human thought.

26 Both the Augusteo symphony orchestra in Rome and Madrid's Orquesta Sinfónica were highly regarded during this period. The Mussolini government razed the former's concert hall in 1936 and the latter suffered during the Spanish Civil War (1936-39).

V

Perhaps it is in astronomy that the most picturesque instances of the use of mathematics have occurred. You all have read of the fall of an apple in a Norfolk garden: how that fall made [Isaac] Newton ponder: and how the scribes (sparing us details perhaps not always known to them) tell us that Newton thence discovered the theory of gravitation.[27] Will you be indulgent if, with only a fragment of calculation and confessedly for an ulterior purpose, I attempt to give you an outline of the process of thought which led to that theory?

Without going back to the beginnings of dynamical science, it will be sufficient to recall one of the experimental observations of Galilei, now almost commonplace enough to be ignored, then novel enough to startle old conceptions, ultimately to destroy some dogmas of theological (I do not say religious) belief, and to recast human ideas of the cosmic universe. He inferred that a dropped stone fell straight down, that is, towards the centre of the earth, because the descent, wherever tried, is always straight down. He also inferred that, in its movement, the stone fell four times as far in the first two seconds as it did in the first, nine times as far in the first three seconds as it did in the first: and so on for the succession of seconds from the beginning: the distance fallen in any number of seconds is measured by the result of taking that number and multiplying it by itself, "squaring it," as we mathematicians call the

27 Voltaire in 1733: "In the like Manner, Pythagoras ow'd the Invention of Musick to the Noise of the Hammer of a Blacksmith. And thus in our Days Sir Isaak Newton walking in his Gardens had the first Thought of his System of Gravitation, upon seeing an Apple falling from a Tree."

calculation.[28] But this only gives a comparative measure among different periods: something more is wanted in order to state what we popularly call an exact measure.

Such a statement demands a knowledge of the exact distance that the stone would fall in a specified time; for example, the exact distance the stone would fall in five seconds from the beginning of its movement. He had noted observations also which answered the demand: the result of calculations (always there is calculation) showed that there is a number, very slightly larger than sixteen (but let me call it sixteen for brevity of statement), which is of crucial importance.

Return now to the measure of the distance through which the stone falls. We saw that the distance is, first of all, measured by a number which is the square of the number of seconds (the number of seconds multiplied by itself). We take that square number, which gives the first and the sort of relative estimate of the fall, and we multiply that square number by the crucial number sixteen, no matter what the square number may have been. The product—what a schoolboy calls the "answer"—is the exact distance through which the stone has fallen in the specified time, when the distance is measured in English feet. Thus in three seconds the distance would be 144 ft.: for the square of three (that is, three multiplied by three) is 9, and the result of multiplying 9 by the crucial number 16 is 144. In five seconds the fall would be 400 ft.: for the square of five (that is, five multiplied by five) is 25, and the result of multiplying 25 by the crucial number 16 is 400. And so on. Let me ask you to remember that the fall of the stone is measured by the square of the number of seconds in the duration

---

28 *Dialogues Concerning the Two New Sciences* (1638): "I begin by saying that a heavy body has an inherent tendency to move with a constantly and uniformly accelerated motion toward the common center of gravity, that is, toward the center of our earth, so that during equal intervals of time it receives equal increments of momentum and velocity." Theorem II, Proposition II: "The spaces described by a body falling from rest with a uniformly accelerated motion are to each other as the squares of the time-intervals em- ployed in traversing these distances."

of the fall: above all, to remember that there is a crucial number sixteen.[29]

Now an explanation of the movements of the planets was being attained, though very gradually. The Copernican opinion, that they moved round the sun as a fixed body, was making its way, because the more it was tested the more continually did it accord with facts.[30] The observations of Tycho Brahe had been reduced to systematic results by the industry of [Johannes] Kepler, who found that planets move almost in circles, actually in oval curves (mathematicians call them ellipses), with the sun in the supreme position at the focus: and these ovals differ from circles only very slightly.[31] But that discovery of Kepler's was only descriptive, corresponding to the description of a stone as falling straight down: it was no sort of measure of the movement of a planet.

Such a measure came, once more, from Kepler's labours: he established two great inferences, laws as they are called. One law was concerned with the constant rate of speed round the sun, as measured by the area swept out by an imagined line drawn from the sun to the planet, thus giving one sort of measure.[32] The other law was concerned with a relation between the size of the oval described by the planet round the sun, and the time which the planet required to effect one complete revolution round the sun: thus giving a second sort of measure.[33] Knowledge had passed out of the describing stage into the measuring stage.

The genius of [Isaac] Newton comes upon the scene.

---

29 Galileo did not know that the acceleration due to gravity on Earth is approximately 32.2 feet per second-squared, half of which is 16.1 feet per second-squared.

30 *On the Revolutions of the Heavenly Spheres* (1543): "Thus indeed, as though seated on a royal throne, the sun governs the family of planets revolving around it."

31 First Law of Planetary Motion, proposed in *Astronomia Nova* (1609).

32 Second Law of Planetary Motion, also proposed in *Astronomia Nova* (1609).

33 Third Law of Planetary Motion or Law of Harmonies, proposed in *Harmonices Mundi* (1619).

According to the chronicles in summary, he saw an apple fall: it made him think of the moon; and thought led him to the theory of universal gravitation. Perhaps the sequence of ideas, a falling apple, a steady moon, and an abstract theory, does not seem very coherent; and it sounds like an extract from *Alice in Wonderland* to be told that the abstract theory can account for the fall of the apple and the steadiness of the moon. Yet it is true, and the explanation comes in this wise.[34]

34 Lewis Carroll (1832-1889), British mathematician, published *Alice's Adventures in Wonderland* in 1865.

VI

The Galilei stone and the Newton apple both fall straight down, that is, to the centre of the earth, supposed to be a globe: that qualitative result happens wherever the fall takes place; it is reasonable to suppose that the fall is due to the earth. There had been another discovery of Galilei's which, at first, had shocked even intellectual belief: stones of different weights and kinds and sizes fall through the same height in the same time; and so it is reasonable to suppose that, in the fall of a stone, the earth is the principal agent, perhaps the only agent, that leads to the fall of any stone anywhere through any height.[35] It is true that the distances which can actually be observed for falling stones are ludicrously small compared with the size of the earth; and the crucial number seems to be a lonely fact, unrelated to anything except a falling stone, unexplained, so far inexplicable. But everything connected with the observation and with the calculations remains unaffected if a hypothesis be propounded under which the earth attracts the stone and is the sole cause of its fall: and there is no other observation, reserved in a background, to render that hypothesis open to objection.

Next, the moon goes round the earth steadily: so far as that movement is concerned, the moon can be declared (always

35 Law of Falling Bodies, *Dialogues Concerning the Two New Sciences* (1638): "Let me once more explain that the variation of speed observed in bodies of different specific gravities is not caused by the difference of specific gravity but depends upon external circumstances and, in particular, upon the resistance of the medium, so that if this is removed all bodies would fall with the same velocity..."

descriptively, and with sufficient accuracy) to move round the earth in a circle with a steady speed.[36] But the stone falls straight to the earth, always with an increasing speed; the moon moves round the earth, always with a steady speed; what earthly connexion can there be between the two movements?

It is here that the hypothetical attracting faculty of the earth enters, through another discovery which had been made by Galilei and formulated precisely by Newton.[37] Under this law a movable body, if subjected to no influence of any kind, and if it moves at all, will only move in a straight line, and only at an unchanging speed in that line; and variation from the straightness of path, or variation of speed even in a straight path, can be effected only by outside influence. Our falling stone moves in a straight path downwards, but with an ever-gathering speed; this change of speed is due to some influence: all falling stones make in a direction towards the very centre of the earth; can the influence, affecting the speed, be due to the earth attracting the stone down to itself? Our moon does not move in a straight line; the deviation from straightness must be due to some influence; we must, for the moment, leave the uniform speed out of count, for the law makes no declaration concerning uniform speed not along a straight line.

Can this influence upon the moon be due to the earth for ever attracting the moon out of the straight line which would be an uninfluenced path, drawing the moon to itself, with just

---

36 The moon's orbital velocity is approximately 1 kilometer per second, according to NASA. On a scale of 0 to 1, where 0 is a circle and 1 is a parabola, the moon's orbital "eccentricity" around Earth is 0.0549, making it nearly a perfect circle. Earth's orbit around the sun is even closer to circular: 0.0167. The orbit of Halley's Comet around the sun is nearly a parabola: 0.967.

37 An early version of the Law of Inertia, in Galileo, *Second Letter on Sunspots* (1613): "And thus a ship, for example, having received one single time some impetus, would move continuously through a quiet sea without ever stopping, and if one were to bring it gently to rest it would remain at rest, provided that in the first case all the extrinsic impediments could be removed, and in the second case no external mobile cause came upon it."

enough influence to keep it at a steadily unchanging distance from itself? On the hypothesis of an attracting quality in the earth, the said quality must explain the deviation of the moon from a straight path, while it can explain the changing speed of the falling stone. The hypothesis must be tested; it is not to be declared true because it has not been disproved.

Critical tests are ready in the observed facts and in the inferences from the facts. But how are the facts to be used? Simply and solely by the application of mathematics, never constructed in this direction by any man before Newton, now the possession of any reasonably capable mathematical undergraduate.[38]

Remember that we have the crucial number, sixteen, for Galilei's falling stone: let us now pay some attention to the moon. As the moon moves steadily in a circle round the earth, in a period of 29 days roughly, and as the distance of the moon from the earth otherwise is known to be approximately sixty times the radius of the earth, the steady speed of the moon in her circle is easily calculated: the inferred estimate will, of course, be based upon the estimated size of the radius of the earth.[39]

An entirely different calculation gives our estimate of the pull to be exercised upon the moon, to drag her regularly and always into the circle away from the straight line which would be her path if there were no influence. Had there been no influence she would have gone along a straight line, forever moving away from the earth farther and farther: as things are, she remains at a steady distance that on the average remains unchanged. Consequently the difference, say at the end of one second, between her actual distance from the earth and what would have been her distance had she moved along a straight line, is the distance through which, during that second, the supposed effective

38 See Appendix: Newton's Moon Test Error.

39 The synodic period, all the phases of the Moon, is about 29.5 days. However, in *Principia*, Newton uses the sidereal period, one revolution of the moon around the earth: 27 days, 7 hours, 43 minutes, or 39,343 minutes total.

influence would have pulled her. That distance is a measure of the influence: but how is the influence to be estimated?

The estimate manifestly cannot come in the same way as for the falling stone. For the stone, only a few hundred feet at the utmost came to account.[40] The moon is 240,000 miles away from the earth and remains at that distance. An assumption that the measure of the attracting influence of the earth is the same for all stones falling upon its surface causes no mental hesitation: the few hundred feet, which are the utmost traversed by a stone, are such a triviality in magnitude compared with the four-thousand-mile radius of the earth that there is no real opportunity for modification of the influence of the earth. But what is to be the assumption made for an attracted moon, if it is attracted? In thought we have to go far to reach the moon: we may as well be bold, go out boldly in thought into the vast space of the solar system, and use the knowledge provided by the Kepler laws which, be it noted, are not hypotheses but are the systematic expression of observed facts.

An easy mathematical calculation indicates the character of the supposed terrestrial influence which does no more than keep the moon going round in a circle at an unchanging distance, but does not bring the moon any nearer as it brings a stone nearer. This calculation uses the Kepler law establishing the relation between the size of a planetary path and the time of revolution along the whole path, and the result of the calculation is to show that the attracting influence of the earth, if it is to be the controlling cause of the movement, must decrease as the square of the number which represents the distance of the attracted body from the centre of the earth. The distance of the moon from the earth is, roughly, sixty times the radius of the earth; and the square of the number 60 is 3,600.[41] Therefore, according to the result of the calculation, the attracting influence of the earth at the distant

---

40 The Leaning Tower of Pisa is about 185 feet tall.

41 NASA: 238,855 miles (average Earth-Moon distance) / 3,959 miles (Earth's radius) = 60.3.

moon is only one 3,600th part of its attracting influence upon a stone falling at its surface.

Now we know that the earth's influence is the same for all falling stones and, ultimately, is measured by the crucial number sixteen. We want to know whether the earth's attracting influence, if any, upon the moon can be the same in kind as upon the falling stone, while, of course, its magnitude will be diminished in proper proportion owing to the difference in distance. Let us denote the crucial number, connected solely with the attracting power of the earth, by the symbol *x*, so dear to mathematicians. For a falling stone at the surface of the earth this number *x* is known to be sixteen. If our hypothesis, that the earth's attracting influence is the sole cause dominating the movement of the moon, can be justified, the number at the distance of the moon would be , just as *x* is sixteen at the surface of the earth. We therefore have to calculate this number for the moon, in the same way as 16 is the calculated number for the stone. When it has been found for the moon, we at once have the value of *x* by a simple operation; and this value of *x* is as crucial for the moon's motion under our hypothesis as is the number 16 for the falling stone. If this crucial number *x* for the moon is inferred from the calculation to be the same as the crucial number 16 for the falling stone, the hypothesis concerning the attracting influence of the earth is so far justified. If it is not, the moon sails on, serenely unexplained. The crucial number for the moon must be found.

Here, Newton was at the crisis of his investigation. He had his facts: there were the established laws, expressing accumulations of observations: there were the old measurements in the accepted units, the foot as the precise unit for distance in the case of the falling stone, the radius of the earth to be expressed in miles as a less precise unit in estimating the distance of the moon. Newton was now to test his theory in the furnace of fact. Everything depended upon the crucial number: would it come out, for the moon, the same as it was for the stone, the critical sixteen?

When this crucial number first emerged from Newton's calculations about the moon, it turned out to amount to something very slightly less than fourteen—let us call it fourteen, as approximately as the critical sixteen. Certainly the had been found to be the number sixteen for the stone, which was as surely sixteen as human observation could make it. The theory was not established as a working hypothesis. Unless unknown and apparently undiscoverable errors had crept in, of such a kind as to vitiate the calculation, the theory could not be maintained. Something must be wrong somewhere: hardly imaginable in the facts or in the data; perhaps—a reluctant perhaps—in the assumption that the earth was the attracting influence upon the stone and the moon. At that stage there was nothing to be done by a mind and a temperament such as were Newton's: he kept silence, consigned his soul not to perdition but to patience, and proceeded to ponder over other riddles of the physical universe.[42]

42 "Not to perdition but to patience." Possible allusion to Hebrews 10:36-39.

VII

Now what is the secret of the situation? It is implied in what has already been said, as though it were a clue in a cross-word puzzle. The calculations involved, through the moon's distance, an estimate of the radius of the earth in miles; and in Newton's day the mile, a geographical or nautical mile, was the sixtieth part of a degree. The calculations involved, through the distances traversed by the stone, the adoption of the foot as the unit. The two different units of length had to be brought into relation. We, of course, take 5,280 feet as equivalent to a mile; so did Newton. But that mile of 5,280 feet is not a geographical mile, the sixtieth part of a degree of latitude; the geographical mile survives solely in one usage, the maritime knot of 6,080 feet, and Newton did not know the fact. A correct estimate had, it is true, been published some thirty years earlier by a London schoolmaster based upon his own observations during a holiday tramp from London to York; but in the Stuart days of the first [King] Charles England was more concerned with political turmoil than with a pedagogue's estimate of a mile, and the schoolmaster's contribution to knowledge remained unnoticed and unknown.[43]

Some thirteen years passed in silence for Newton. In 1679 a friend told him of a recent French research in geodesy which gave a corrected estimate of the length of a degree of latitude: still, of course, the sixty geographical miles, but a little over sixty-nine

43 The Stuart dynasty ruled England from 1603 to 1714, interrupted by the English Civil War, the Interregnum, and the rule of William and Mary of Orange. The "schoolmaster" is Richard Norwood (1590-1676).

of the customary civic miles of 5,280 feet.[44] Thus one datum in Newton's calculation had been wrong: he had used a wrong number, 60, instead of a right number, 69. In the light of this new knowledge Newton returned to his old calculations: the use of the more accurate measure of the radius of the earth brought out the crucial number for the moon to be in exact and complete agreement with the crucial number 16 for a falling stone. His hypothesis concerning the attracting influence of the earth was verified. It became a trustworthy working hypothesis which, in due course of its development, though then he knew it not, was to achieve a revolution in thought as regards the human knowledge of the physical cosmos.

But his mathematical triumph was only the beginning of new mathematical work. Suddenly there loomed before him a portentous difficulty. All his calculations had treated the moon, the earth, the sun, as points—points which, in Euclid's definition[45] familiar to Newton, had no parts and no magnitude: an earth of no magnitude; a sun of no magnitude; truly portentous assumptions! To Newton's mind a fresh difficulty was only a fresh problem to be solved: and, here, he made another mathematical discovery which was of supreme importance for his theory. The moon, the earth, the sun, the planets, all of them round bodies, are very nearly spherical in shape; and Newton's fresh discovery was that the complete attraction of a sphere on an outside body was the same, in all respects, as if the whole of the matter were condensed at the centre.[46] In fact, his mathematics showed that he could regard the moon, the earth, the sun, as attracting points.

---

44 Geodesy: the science of measuring Earth. The French researcher is Jean Picard (1620-1682). The "friend" who mentioned Picard's results to Newton is Robert Hooke (1635-1703), Newton's professional rival.

45 *Elements*, Book I, Definition I: "A point is that which has no part."

46 *Principia*, Book I, Section XII, Proposition LXXV, Theorem XXXV: "If to the several points of a given sphere there tend equal centripetal forces decreasing in a duplicate ratio of the distances from the points; I say, that another similar sphere will be attracted by it with a force reciprocally proportional to the square of the distance of the centres."

The difficulty disappeared: the theory could stand. Now he could go forward, to test the theory further by further extension: the result was his theory of universal gravitation according to which every particle of matter in the universe attracts every other particle[47] according to a definite law: the attraction was measured by dividing one fixed number by a number which represented the square of the distance between the attracting particles.[48] He worked out the theory in dctail, under this law; the calculated results everywhere agreed with the observed phenomena; and the result was the publication, in 1687, of his *Mathematical Principles of Natural Philosophy* [better known as the *Principia*], one of the great books of the world.

It was an achievement made through mathematics by a mathematician. It has been confirmed and amplified since Newton's day by generations of mathematicians of all the ages and all climes: by means of it, and still using their mathematics, they have made predictions that are daily verified in the movement of the planets; they have predicted the return of comets, they discovered a new planet, and, passing from fact as commonplace as the creation of the *Nautical Almanac* to the shadowy realms of high speculation, they measure the bounds of the universe, assign a limit in the past to its evolution from chaos, nay, they will even assign a limit to its future—a very remote limit, happily for our sense of ease—when it may subside into an exhausted extinction.[49] Personally, I prefer fact to speculation.

---

47 *Principia*, Book III, Proposition VII, Theorem VII: "That there is a power of gravity tending to all bodies, proportional to the several quantities of matter which they contain."

48 The inverse-square law of universal gravitation. For example, *Principia*, Book III, Proposition VIII, Theorem VIII: "In two spheres mutually gravitating each towards the other, if the matter in places on all sides round about and equi-distant from the centres is similar, the weight of either sphere towards the other will be reciprocally as the square of the distance between their centres."

49 *Encyclopedia Britannica* (1926) reported 18 periodic comets with well-understood orbits. Today, the International Astronomical Union's Minor Planet Center lists nearly 500. *Nautical Almanac*: An aid to navigation

published from 1766 onward by the British government (now jointly with the Americans). The universe is presently believed to be 13.8 billion years old—with an extremely unclear expiration date. The term "exhausted extinction" may allude to the "heat death" theory proposed by Lord Kelvin (ca. 1862), suggesting that the sun or even the universe would eventually dissipate all of its energy. Major advances in cosmology were made in the years immediately following Forsyth's lecture, including Edwin Hubble's observations of redshift that indicated the universe was expanding (1929) and Georges Lemaître's theory of the "primeval atom" (1931) that presaged the Big Bang theory.

VIII

Behind my account of the emergence of the theory of universal gravitation, there stands a purpose of warning rather than edification. Often in recent years you will have heard the name of [Albert] Einstein: the journals and newspapers appeared to palpitate with excitement over his discovery, which was very real and of real importance: but there always was, and there remains, a grave dearth of comprehensible explanation as to the gist of his discovery. The writers found the word relativity: it used to bear a philosophical significance mainly connected with a doctrine that existence and observation are not independent of the living and observing mind. The association with the physical universe was rather nebulous in explanation: that mattered little; was not Einstein's important (but dimly understood) discovery connected with relativity? So, imitating the image-makers at Ephesus in the days of the apostle Paul, the expositors proclaimed that "Great is relativity."[50] Here are the facts, so far as they are concerned with the theory of gravitation.

That theory of gravitation was found to embrace all astronomical observations. By the assumption of that theory as a working basis, Halley in 1706 had predicted the periodic return of a particular comet: it duly verified the prediction in 1758, in 1835, in 1910; and its next coming is due only fifty-eight years hence. By the assumption of that theory [John Couch] Adams

50 "Great is Diana of the Ephesians." The New Testament records the visit of Christian missionaries (affiliated with Saint Paul) to the ancient Greek city of Ephesus, the ruins of which are located in present-day Turkey. Ephesus was home to the Temple of Diana (or Artemis), one of the Ancient Wonders of the World. See Acts 19:24-28.

and [Urbain] Leverrier, independently, through a vast mathematical calculation applied to some seeming irregularities of the planet Uranus, discovered the existence of a still more distant planet Neptune, never observed before the discovery made in their studies. The mathematical genius of [Pierre-Simon] Laplace had set the phenomena of the astronomical heavens in mathematical order. Everything was settled into its place, or almost everything.

There remained one or two minute somethings. When the most elaborate calculations had been completed, there remained a little unexplained fluctuation in the movement of the moon, so often accused of fickleness by the poets; and one planet, fitly named after Mercury, the not over-righteous messenger of the gods in classical mythology, was found to be acting in a manner inconsistent with strict mathematical propriety. The vagaries of Mercury were very slight: its orbit is far more tilted than that of any other planet, and is more swiftly traversed than that of any other planet; but the vagaries could not be explained by tilt or speed, and their source (that is, the cause of the difference between what is actually seen and what calculation led astronomers to expect) could not be discovered. The irregularities made no difference to daily life: they made no other difference to astronomical life: but there they were. Nature would not change so as to conform to man's explanation, so man's explanation must be examined in order to be brought into conformity with nature. The accuracy of the calculations could not humanly be challenged, for they obeyed every test that human reason could devise. Perhaps there was something, an extremely tiny something somewhere (for, remember, the deviations are extremely slight), something wrong in the fundamental assumptions upon which the calculations were based. And here is an illustration of the difference between practice and thought: "to stop that nonsense" an engineer would have driven a rivet, would have inserted a guiding slot, would have added a tightening chain round the dome of St. Paul's [Cathedral]. But rivets and slots and chains are devastations in thought, not remedies: so the fundamental assumptions must be revised. It had been assumed that the solar system is fixed, absolutely fixed, in the universe: is that dogma true? Even if it is not true, the alternative of a wandering solar system would hardly explain the minute domestic vagaries of one rather insignificant member.

It had been assumed that the measurement of the same thing, in varying circumstances and at diverse times, made as accurate as human skill can achieve, and corrected for every incidental error that human skill can detect, is always the same, whatever be the time and the place of the thing measured and the measurement made. The dogma seems an obvious inference from what we are pleased to call common sense: like so many obvious things, it cannot be proved and it cannot be submitted to adequate test: in fact, common sense continues to call it obvious. But how if it be not true? Your standard of measurement, used to measure a magnitude in passing time and on a moving body, would change in the same way (if at all) as the measured magnitude, and you could not detect the change, if there be change: and, in that case, your measures would no longer be absolute: they would be relative, perhaps to time, perhaps to place or change of place, perhaps to the way of change. Nothing remains absolute: yet that was a fundamental assumption, stated explicitly, in the Newtonian philosophy.

Now this notion of relativity is not new: it did not come into scientific thought with Einstein: it was not unknown to Newton himself: and men have pondered over it often in the last fifty years. But there is the customary difficulty: an idea may come into the range of thought: how can it be garbed in expression, for discussion, for calculation? You will remember the wisdom of a character in George Eliot's *Theophrastus Such*, "There's some as thinks one thing: there's some as thinks another thing: but my opinion's different."[51] Men, charging themselves with the

51 This quotation is slightly garbled. The character Pummel states: "Many gives their opinion, but if I was to give mine, it 'ud be different." Eliot was

formulation of relativity, have devised various modes of its mathematical expression; and, so far as concerns astronomy, there is not the unanimity of expression which, if there, would be a rudimentary recommendation.

It was the work of Einstein which was the first to explain—or, at any rate, to account for—the behaviour of Mercury: and therefore it is to Einstein that the glory rightly accrues. The method of Einstein, however, is not the sole one that gives an explanation; for the mathematical methods of other men have also led to that result, after Einstein.[52] But neither the method of Einstein nor the method of any other mathematician has yet succeeded in explaining the slight caprice of the moon.[53]

How far the popular expositors had made themselves acquainted with the investigations in all accessible issues, before some of them tried to make our flesh creep with their pronouncement upon the passing of gravitation, might be the curious quest of some doubting scientific Thomas Didymus.[54] But of one thing we may be reasonably sure: neither in that theory, nor in any other living theory, has the last word been said. For the last words of all are the epitaph on a tomb: and the Newtonian theory is still alive sufficiently to provide man with a working hypothesis of the natural universe in which he lives.

---

the *nom de plume* of Marian Evans (1819-1880), a prolific English writer. Theophrastus was also a 4th century B.C. student of Aristotle.

52 Likely reference to prominent German astronomer Hugo von Seeliger (1849-1924), who argued that Mercury's misbehavior could be due to cosmic dust near the sun. His theory was later refuted.

53 Ernest William Brown (1866-1938) proposed in 1926 that the moon's "slight caprice" was actually due to a slight caprice of Earth's rotation: "The answer to a question sometimes asked as to whether the Einstein effect might be responsible is clearly in the negative. One reason is that it is too small." Later work proved Brown correct.

54 Saint Thomas the Apostle doubted Christ's resurrection. See John 20:24-25. Thomas (Aramic-Hebrew for "twin") Didymus (Greek for "twin") literally translates to "Twin Twin."

IX

Before I close, there are two inferences which I would appeal to you not to draw from my lecture. It is not the fact that the services of mathematics are restricted to the subjects that have been selected for mention or illustration. It is not the fact that the science of mathematics exists or thrives, solely or mainly in order to find or to further applications in other sciences.

The range of human activities, within which mathematics and the ideas of mathematics are called upon or are utilized for service, is so vast and so varied that I can do no more, at this stage, than mention the names of some of the regions where the science of mathematics is active. Enough has been said on the score of astronomy. In engineering of every type, ranging from the ancient ship of the sea to the modern ship of the air, in your buildings, in your bridges, in your tunnels, its help is indispensable.

Is it an occupation so sedate as book-keeping, say by double entry? That process is but an iterated application of the use of the plus and minus signs of algebra, though the book-keeper is not usually aware of the fact.

Is it a science so relatively modern as physical chemistry? Much of that science is based upon the mathematical expression of some of the laws of change in physical nature.

Is it a science which apparently is so far remote from the usual conception of mathematics as is physiology, the science of the processes of living organisms? That human science is examining the phenomena of the activities of such organisms, is expressing quantitatively those activities, is dealing mathematically with those quantitative expressions: there are the researches on the expenditure of muscular energy, the researches on industrial fatigue. Did not a sectional president startle the British Association the other day by hinting (though without adequate justification) that a distinguished physiologist, trained as a mathematician and using his knowledge, was on the verge of discovering the mystery of life itself?[55]

Is it the spotting of invisible guns on a far-flung battlefield? My friend and former colleague, your worthy professor of mathematics[56] in this College, could tell you of the combined efforts of mathematicians and physicists who, by means of instruments and calculations connected with the properties of sound, were able to devise means of approximate location. And those same properties were used, in connexion with mathematical instruments constructed for the purpose, to detect, from the surface of the sea, not merely the fact, but the direction of approach, of a hostile submarine.

The citations of utility could be continued almost without limit: only one other subject will be indicated. If there is anything, that appeals to the average inhabitant of these islands and is a frequent topic of unfruitful conversation, it is the weather: and a feature of our morning newspaper is the forecast issued by the Meteorological Office.[57] But the occupants of that office are not like the ancient soothsayers, waiting for inspiration from gifts tendered and from sacrifices rendered. They collect scattered information, co-ordinate that information by settled mathematical processes, draw their mathematical diagrams to represent that co-ordinated information, and, inferring tendencies from movements indicated by their diagrams, issue their calculated forecast.

---

55 See Appendix: Frederick Donnan Controversy.

56 Forsyth's original footnote here reads: "(Lieutenant-Colonel) Professor A. R. Richardson, D.S.O." See Appendix: Archibald Read Richardson.

57 Established in 1854.

All these—there are more that crowd into the mind—may be an adequate recital showing how essential and how extensive is the service rendered by the rather silent science of mathematics to the practical life of mankind.

X

Most of all, I would not wish to have you possessed by the notion that the pursuit of mathematics by human thought must be justified by its utility for practical uses in life. Such a justification would, it is true, require the inclusion of mathematics in the scheme of any university where technical studies are deemed important; and in the growing prosperity of your University College, the more generously you open your technical courses to mathematics, even in their utilitarian aspect, the more will the professional students ultimately benefit.

But having paid tribute to the demands of the market-place, let me speak for mathematics as a pure science of progressive knowledge, worthy to claim the devotion of the finest intellects to its pursuit, worthy to challenge the respect of man towards the world of learning. One of the high ideals of mankind through all the ages, and in all civilizations, has inspired the search for more knowledge wherever it can be found or be attained. Ever since man has attempted to acquire ordered knowledge, the science that deals with number and deals with form has been pursued for its own sake because, thereby, the human spirit can find unending satisfaction and unending occupation. And the creations of mathematical science have been the glory of the nations.

We may recall the lost dominion of Babylon, we may think of her hanging gardens as, in their time, a wonder of the world; we may sigh over her pomp and her luxury, gone like a dream in the visions of the night; but the contributions of her Chaldean priests to astronomy survive to this day.[58] Greece has an immortal name from her art, from her literature, from her philosophy: her rival schools of Plato and of Aristotle still dominate the Western world; but her fame is no less immortal by her bequest of geometry and of number, two of the purest theories ever devised by the human mind. We may hold Semitic science in low regard: but Arabic learning kept science alive when the rest of the intellectual world was torpid; it created algebra as the marvellous extension of the Greek arithmetic, and gave us the very digits we use and the scale of ten, so familiar to us as to seem part of our existence: we can pay our mental homage of remembrance to those ancient Arabs in southern Spain.[59]

What the modern peoples have thus inherited they have amplified beyond recognition; and the nations of the West—France, Germany, Italy, America across the ocean, our own people—have lived in an unending lifegiving rivalry in the creation of mathematical knowledge, sought for its own sake, its domain as boundless as human thought itself.

For there is progress still in mathematical science; there will always be progress of increasing knowledge in a world that is not dead. Results have been achieved by the noble army of great spirits of the past, and their achievements are the possession of the living. But those very achievements are the stimulus to the living that they, in their turn, shall endeavour to advance knowledge. And this pursuit is to be made by the living spirits for the

58 The Seven Wonders of the Ancient World included the Colossus of Rhodes, Great Pyramid of Giza, Hanging Gardens of Babylon, Alexandrian Lighthouse, Mausoleum at Halicarnassus, Olympian Statue of Zeus, and Ephesian Temple of Artemis (or Diana). The location (and existence) of the Babylonian gardens is disputed. Except for the Great Pyramid, all of the wonders have been destroyed.

59 Semitic: Hebrew, Arabic. Mathematician al-Khwarizmi (ca. A.D. 800) is the etymological godfather of the term "algorithm" and authored the treatise *al-Jabr*, as in "algebra." Though a Persian, he worked in Baghdad. The scale of ten is typically referred to as the Hindu-Arabic number system, given its intellectual pathway from India across the Middle East to the Mediterranean Sea. Reference to the universities and libraries of Muslim Spain (al-Andalus), typically Toledo and Cordoba.

sake of new knowledge, not for the sake of new glory, not for the sake of new benefit. If utility should come, well and good: but we need trouble no more about immediate utility as an aim than the Greeks troubled about the utility of their conic sections or Newton troubled about the utility of the gravitation theory.[60]

So here, amid this community in a centre of commercial activity, in this home of high learning which has been established for the betterment of men and women as human citizens, let me plead, if pleading be needed, for the highest consideration to be given to the pursuit of pure knowledge as well as technical training, not neglecting mathematics, once called the Queen of the Sciences.[61] The wind bloweth where it listeth, and the spirit of knowledge does not follow the quest for wealth and power; but the creation of new knowledge makes for the high repute of a nation, alike in the days when its influence is dominant and in the more distant days when its doings shall have been recorded on the scroll of time.[62]

---

60 Conic sections are the shapes that result from slicing a cone: circle, ellipse, parabola, and hyperbola. Gravity explains the elliptical orbits of the planets.

61 According to his biographer, Carl Fredrich Gauss (1777-1855) described mathematics as "die Königin der Wissenschaften," Queen of the Sciences.

62 "The wind bloweth where it listeth." John 3:8. Henry Kirke White (1855):

> As o'er the scroll of Time I pore,
> And safely spell of ancient lore,
> Till I can rightly guess of all
> That Plato could to memory call,
> And scan the formless views of things;
> Or with old Egypt's fettered kings,
> Arrange the mystic trains that shine
> In night's high philosophic mine;
> And to thy name shall e'er belong
> The honors of undying song.

Mathematics as Daimonic Science

KEVIN M. KAMBO

Historically, the Greek term *daimon* has referred to higher, intelligent powers that affect, for good or for ill, the lives of human beings. When those forces are benevolent, we think of protective spirits and guardian angels, acting similarly to how Athena and Hermes watch over Homer's Odysseus; when the forces are inimical, formidable threats to the peace we all seek, we speak of demons. Such, for example, are Loki, the Norse god, or the devil, whom Christians view as humanity's chief accuser and adversary. Andrew Forsyth's lecture invites us to attend to the promise and peril of mathematics as one such daimonic influence.

Although Forsyth's purpose is to articulate the noble offices and uses of mathematics, and thereby to inspire appreciation, his lecture is delivered in a poignant context: the end of 1928, roughly a decade after the First World War and a decade before the Second. For all the many conveniences and amenities by which mathematics ever makes us more powerful in preserving and improving life, it is continually marshalled to endow us with yet more dreadful deathworks: machine guns, gas weapons, flamethrowers, extermination camps, the atomic bomb. Francis Bacon and René Descartes might have prophesied that mathematical science would deliver us the amelioration of man's estate, but our mastery over nature also brought Oppenheimer to condemn himself in the words of the *Bhagavad Gita*: "Now I am become

Death, the destroyer of worlds." Twentieth Century mathematics delivered into human hands a sword such as to rival the legendary stroke that laid low the firstborns of Egypt and the Assyrians at Sennacherib. Daimons, we recall, are guides and guardians, but also deceivers and destroyers. That said, Forsyth's mind is very much on the former role, and his words are still worth our attention because they retain (and likely have surpassed) their original relevance, almost a century after he offered them.

The title of the essay, "Mathematics in Life and Thought," invites us to consider the place of this guiding science in our personal and political lives. The distinction between Life and Thought is not technical—there is no thought, Plato writes, without life—and so I shall chart Forsyth's meaning by employing a similar paradigm, inherited from the Cappadocian theologian Gregory of Nyssa. Gregory divides human activity into three categories—action, speech and thought—which categories are useful for surveying how Forsyth lays out the guidance, protection and excellences that mathematics offers us.

In Plato's *Republic*, Socrates proposes four numerical sciences—arithmetic, geometry, astronomy and harmonics—as preparatory to philosophy. These sciences liberate the mind from overreliance on our physical senses, but Socrates notes that they nevertheless have material utility in mundane life. For example, arithmetic is useful for household economics, and geometry is employed by military tacticians studying terrain. In similar vein, Forsyth notes that mathematical operations are ubiquitous in daily life, even though very few people understand the mathematical principles and proofs supporting and making reliable those operations. The peasant in the market, the clerk in the office, the shopkeeper and the scheduler—in brief, the vast diversity of ordinary professionals who testify as a great cloud of witnesses—have been so comprehensively habituated to carry out calculations that they barely recognize that the invisible order in which their commerce moves, lives and has its being is a mathematical order. And what is true of the individual is arguably even more fundamental politically.

The centrality of mathematics to our markets and economies is obvious. More remarkable, I wager, is Forsyth's discussion of forecasting. He notes that it is through the ministrations of mathematics that weather prediction has been delivered from the province of soothsayers and borne into the instruments of meteorologists. Our prophetic scientists "collect scattered information, co-ordinate that information by settled mathematical processes, draw their mathematical diagrams to represent that co-ordinated information, and, inferring tendencies from movements indicated by their diagrams, issue their calculated forecast." In Forsyth's time, science had compassed the terrestrial elements: earth, fire, wind and water. In these latter days, forecasting has overtaken humanity, the fifth element. Little does the ordinary person appreciate how much of our activities are guided and driven by mathematical modelling and prediction. No matter the domain—whether sport, politics, medicine, entertainment, travel, marketing or insurance—we scarcely notice the algorithms that, like vivifying spirits over the surface of the waters, hover around us. So much, then, for the world of action. Let us briefly turn to speech and thought.

Socrates' daimon exercised a negative office: it chiefly prevented him from doing and saying things he should not. A mathematical sense, or a feel for quantity, plays a similar part in life. Like George Orwell, Forsyth notices and laments mankind's tendency to repeat hearsay thoughtlessly. Many rumors are sustained because too many people lack a mind for magnitude; they repeat the fabulous and fantastical because they are unwilling or unable to reckon (etymologically, *to count*) the mathematical probability of what they hear—and then themselves propagate. Rumor might never be eliminated, but a mathematical sense would at least be one more valuable guardian of our tongues.

As with the tongue, so with thought. Calculation is often a sufficient method for testing and correcting theory. Even more than in Forsyth's day, we deploy mathematics to evaluate scientific models and theories. But purification does not exhaust its guiding role; it also leads. Forsyth shows how, historically, mathematical systemization has revealed new frontiers for intellectual exploration and discovery: mathematics does not just eliminate false possibilities; it also illuminates parts of the map of knowledge previously darkened by the dragons of the ignorant and unimaginative. In guarding, guiding and governing thought, mathematics is thus a ruling science, an imperial discipline on whose empire the sun never sets.

All that said, Forsyth's most ambitious claim is that, beyond all its benefits, mathematics is most purely pursued when enjoyed for its own sake. This achievement cannot be described adequately; it is known firsthand. Take, for instance, the act of grasping the beauty in an elegant proof: one sees only for oneself; only initiates can taste and see the goodness of this achievement. But, a scent of a hint can be caught in witnessing Forsyth revel in the history of science, an object of study largely abandoned in both the scientific and humanistic corridors of contemporary academia. In Forsyth, one witnesses a rare human excellence: pursuit of the knowable, for its own sake. In mathematics, this pursuit raises the human mind to play and participate in an eternal order, the order that, in Dante's words, moves the sun and the other stars.

References

Archimedes. "The Sand-Reckoner," in *The Works of Archimedes*, edited by Thomas L. Heath (Cambridge University Press, 1897), p. 221-232; p. 222.

Aristotle. *De Caelo* (On the Heavens), translated by J. L. Stocks (Clarendon Press, 1922), Book II, Paragraph 3, Lines 13-21, p. 286a.

Brown, Ernest. "The Evidence for Changes in the Rate of Rotation of the Earth and their Geophysical Consequences," *Proceedings of the National Academy of Sciences*, Vol. 12 (June 1926), pp. 406-413.

Copernicus, Nicolaus. *On the Revolutions of the Heavenly Spheres*, edited by Jerzy Dobryzycki and translated by Edward Rosen (Warsaw-Kraków, 1978), Book, I, Chapter 10, p. 22.

Dante, Alighieri. *The Divine Comedy: Volume 3: Paradise*, translated by Dorothy L. Sayers and Barbara Reynolds (1962), Canto XXIX, Lines 82-87, 94-96.

Eliot, George (Marian Evans). *Impressions of Theophrastus Such* (1879), pp. 121-122.

Faraday, Michael. "Experimental Researches in Electricity," *Philosophical Transactions of the Royal Society of London*, Vol. 122 (1832), pp. 125-162.

Galen. *On the Elements According to Hippocrates*, translated by Phillip de Lacy (Akademie Verlag, 1996), p. 63.

Galileo, Galilei. *Dialogues Concerning Two New Sciences*, translated by Henry Crew and Alfonso de Salvio (Macmillan Company, 1914), pp. 74, 174, available at Wikisource.

—. "Second Letter on Sunspots" (1613), in *On Sunspots*, translated by Eileen Reeves and Albert Van Helden (University of Chicago Press, 2010), p. 125.

Hamilton, William Rowan. "On Some Results of the View of a Characteristic Function in Optics," *Report of the Third Meeting of the British Association for the Advancement of Science* (John Murray and Sons, 1834), pp. 360-370.

Hertz, Heinrich. *Electric Waves*, translated by D. E. Jones (Macmillan and Company, 1893).

Lucretius. *De Rerum Natura*, translated by William Ellery Leonard (E. P. Dutton, 1916), Book I, Lines 500-507, available at Perseus Digital Library.

Maxwell, James Clerk. "A Dynamical Theory of the Electromagnetic Field," *Philosophical Transactions of the Royal Society of London*, Vol. 155 (1865) pp. 459-512.

Newton, Isaac. *Newton's Principia: The Mathematical Principles of Natural Philosophy*, translated by Anderw Motte (Daniel Adee, 1846), available at Wikisource.

Peck, Harry Thurston. "The Woman of To-Day and of To-Morrow," *Cosmopolitan*, Vol. XXVII, No. 2 (June 1899), pp. 149-162; p. 150

Pliny the Elder. *The Natural History*, translated by John Bostock and H. T. Riley (1855), Book XXXV, Chapter 36, available at Perseus Digital Library.

Proclus. *A Commentary on the First Book of Euclid's Elements*, translated by Glenn R. (Morrow (Princeton University Press, 1992), pp. 56-57.

Thomson, George Paget. "The Waves of an Electron," *Nature*, Vol. 122, No. 3069 (August 25, 1928), pp. 279-282.

Thomson, J. J. *The Corpuscular Theory of Matter* (Archibald Constable & Co., 1907), pp. 33-34.

Thomson, William (Lord Kelvin). "On the theory of the electric telegraph," *Proceedings of the Royal Society of London*, Vol. VII (1856), pp. 382-399.

Voltaire. Florence Donnell White, *Voltaire's Essay on Epic Poetry: A Study and an Edition* (Bryn Mawr College, 1915), p. 131.

Von Waltershausen, Wolfgang Sartorius. *Gauss zum Gedächtniss* (Verlag von S. Hirzel, 1856), p. 79.

White, Henry Kirke. "To Contemplation," *The Poetical Works and Remains of Henry Kirke White* (E. H. Butler, 1855), pp. 129-134.

Appendices

Gambling Probabilities

Given a series of n matches, each with 3 possible outcomes, the total number of ways of guessing the outcome of 12 matches is:

$$3^{12} = 531{,}441$$

We can calculate **C**, number of ways 12 matches can be "combined"[63] in sets of r matches:

$$_{12}\mathbf{C}_r = \frac{12!}{r!(12-r)!}$$

Given r matches, the outcome of each of which can be guessed incorrectly 2 ways, the number of possible incorrect outcomes is given by this expression:

$$2^r$$

Given this information, we can say calculate **X**, the number of ways r matches can be guessed incorrectly:

$$\mathbf{X} = \frac{12!}{r!(12-r)!} \times 2^r$$

63 For a brief history of early combinatorics, see David Eugene Smith, *History of Mathematics: Special Topics of Elementary Mathematics*, Vol. II (Ginn and Company, 1925), pp. 524-531.

Finally, we can apply this formula to generate this table of results.

r	$_{12}C_r$	2^r	**X**
1	12	2	24
2	66	4	264
3	220	8	1,760
4	495	16	7,920
5	792	32	25,344
6	924	64	59,136
7	792	128	101,376
8	495	256	126,720
9	220	512	112,640
10	66	1,024	67,584
11	12	2,048	24,576
12	1	4,096	4,096

Cult of Diana

Forsyth refers to "the historic cult in the Arician grove." This is a reference to the ancient Roman cult of Diana centered on Lake Nemi near the town of Aricia and its nearby woods. He may have been familiar with this text:

> In this sacred grove there grew a certain tree round which at any time of the day and probably far into the night a strange figure might be seen to prowl. In his hand he carried a drawn sword, and he kept peering warily about him as if every instant he expected to be set upon by an enemy. He was a priest and a murderer; and the man for whom he looked was sooner or later to murder him and hold the priesthood in his stead. Such was the rule of the sanctuary. A candidate for the priesthood could only succeed to office by slaying the priest, and having slain him, he retained office till he was himself slain by a stronger or a craftier.[64]

For the poem that Forsyth quotes, see Thomas Babbington Macaulay, "The Battle of the Lake Regillus," *Lays of Ancient Rome* (1842):

> From the still glassy lake that sleeps
> Beneath Aricia's trees—
> Those trees in whose dim shadow
> The ghastly priest doth reign,
> The priest who slew the slayer,

64 James Frazer, *The Golden Bough: A Study in Comparative Religion*, Vol. I (1894), p. 2.

And shall himself be slain…

Lake Regillus was the site of an important battle in Rome's republican history. Elsewhere in the text, Forsyth alludes to the visit to Ephesus in the Book of Acts. Ephesus was home to the Temple of Diana (Artemis in the Greek pantheon), one of the Ancient Wonders of the World.

Diameters of the Sun and Earth

In *On the Sizes and Distances of the Sun and Moon*, ancient Greek mathematician Aristarchus proposes a series of hypotheses and propositions concerning the spatial relationships between the Sun, Earth, and Moon. He did not have the tools of trigonometry nor even the knowledge that 360 degrees comprise a circle. Nonetheless, he was able to estimate that the diameter of the Sun is approximately seven times the diameter of the Earth. This is how he did so.[65]

The accompanying chart (Fig. 1) is nearly identical to the chart Aristarchus provides in his text. The only differences are in notation: we will use descriptive lengths rather than points.

To find the lower bound, note Aristarchus proved in an earlier proposition that

$$\frac{\text{Length Shadow Line}}{\text{Sun's Diameter}} < \frac{1}{9} \quad \text{so by inversion} \quad \frac{\text{Sun's Diameter}}{\text{Length Shadow Line}} > \frac{9}{1}$$

Given similar triangles in the chart and by conversion, we can say

$$\frac{\text{Dist. Sun to Vertex}}{\text{Dist. Shadow Line to Vertex}} > \frac{9}{1} \quad \text{and} \quad \frac{\text{Dist. Sun to Vertex}}{\text{Dist. Sun to Shadow Line}} < \frac{9}{8}$$

65 This reconstruction is based on Thomas Heath, *A History of Greek Mathematics: From Aristarchus to Diophantus*, Vol. II (Clarendon Press, 1921), pp. 8-15; Thomas Heath, *Aristarchus of Samos: The Ancient Copernicus* (Clarendon Press, 1913), pp. 403-409; and Albert van Helden, *Measuring the Universe: Cosmic Dimensions from Aristarchus to Halley* (University of Chicago Press, 1985), pp. 6-9. One book stands out from the rest: Christián C. Carman and Rodolfo P. Buzón, *Aristarchus of Samos: On the Sizes and Distances of the Sun and Moon* (Routledge, 2023), pp. 240-243.

In another earlier proposition, Aristarchus proved that

$$\frac{\text{D Sun to Earth}}{\text{D Earth to Moon}} > \frac{18}{1} \quad \text{therefore} \quad \frac{\text{D Sun to Earth}}{\text{D Earth to Shadow Line}} > \frac{18}{1}$$

By inversion, we can say

$$\frac{\text{D Earth to Shadow Line}}{\text{D Sun to Earth}} < \frac{18}{1} \quad \text{so by composition} \quad \frac{\text{Sun to Shadow Line}}{\text{Sun to Earth}} < \frac{19}{18}$$

By equality of terms we can say

$$\frac{\text{D Sun to Vertex}}{\text{D Sun to Earth}} < \frac{9 \times 19 = 171}{8 \times 18 = 144} = \frac{19}{16} \quad \text{so by conversion} \quad \frac{\text{D Sun to Vertex}}{\text{D Earth to Vertex}} > \frac{19}{3}$$

Given similar triangles, we can conclude

$$\frac{\text{Sun's Diameter}}{\text{Earth's Diameter}} > \frac{19}{3} \quad \text{which is the lower bound.}$$

To find the upper bound, note Aristarchus proved in an earlier proposition that

$$\frac{\text{D Earth to Moon}}{\text{D Shadow Line to Moon}} > \frac{675}{1} \quad \text{so by conversion} \quad \frac{\text{D Earth to Moon}}{\text{D Earth to Shadow Line}} < \frac{675}{674}$$

In another earlier proposition, Aristarchus proved that

$$\frac{\text{D Sun to Earth}}{\text{D Earth to Moon}} < \frac{20}{1}$$

Therefore, by equality of terms, we can say

$$\frac{\text{D Sun to Earth}}{\text{D Shadow Line to Moon}} < \frac{675 \times 20 = 13{,}500}{674} = \frac{6{,}750}{337}$$

Further, by inversion and composition, we can say

$$\frac{\text{D Sun to Shadow Line}}{\text{D Sun to Earth}} > \frac{337 + 6{,}750 + 7{,}087}{6{,}750}$$

In yet another earlier proposition, Aristarchus proved that

$$\frac{\text{Length Shadow Line}}{\text{Sun's Diameter}} > \frac{979}{10{,}125} \text{ so by inversion } \frac{\text{Sun's Diameter}}{\text{Length Shadow Line}} > \frac{10{,}125}{979}$$

By similar triangles, we can say

$$\frac{\text{Sun's Diameter}}{\text{Length Shadow Line}} = \frac{\text{D Sun to Vertex}}{\text{D Shadow Line to Vertex}} < \frac{10{,}125}{979}$$

By conversion, we can say

$$\frac{\text{Sun to Vertex}}{\text{Sun to Shadow Line}} > \frac{10{,}125}{9{,}146} \text{ but } \frac{\text{Sun to Shadow Line}}{\text{Sun to Earth}} > \frac{7{,}087}{6{,}750}$$

Therefore, by equality of terms, we can say

$$\frac{\text{Sun to Vertex}}{\text{Sun to Earth}} > \frac{10{,}125 \times 7{,}087 = 71{,}755{,}875}{9{,}146 \times 6{,}750 = 61{,}735{,}500} > \frac{43}{37}$$

By conversion, we can say

$$\frac{\text{D Sun to Vertex}}{\text{D Earth to Vertex}} < \frac{43}{6}$$

Given similar triangles, we can resolve the upper bound:

$$\frac{\text{Sun's Diameter}}{\text{Earth's Diameter}} < \frac{43}{6}$$

Therefore, the ratio bounded by the inequalities is

$$\frac{19}{3} < \frac{\text{Sun's Diameter}}{\text{Earth's Diameter}} < \frac{43}{6}$$

Aristarchus has thus proved that the Sun's diameter is approximately seven times the diameter of Earth, if all of his previous assumptions and propositions are correct. Given his limitations, it should not be surprising that some were wrong, yielding a wildly inaccurate estimate here.

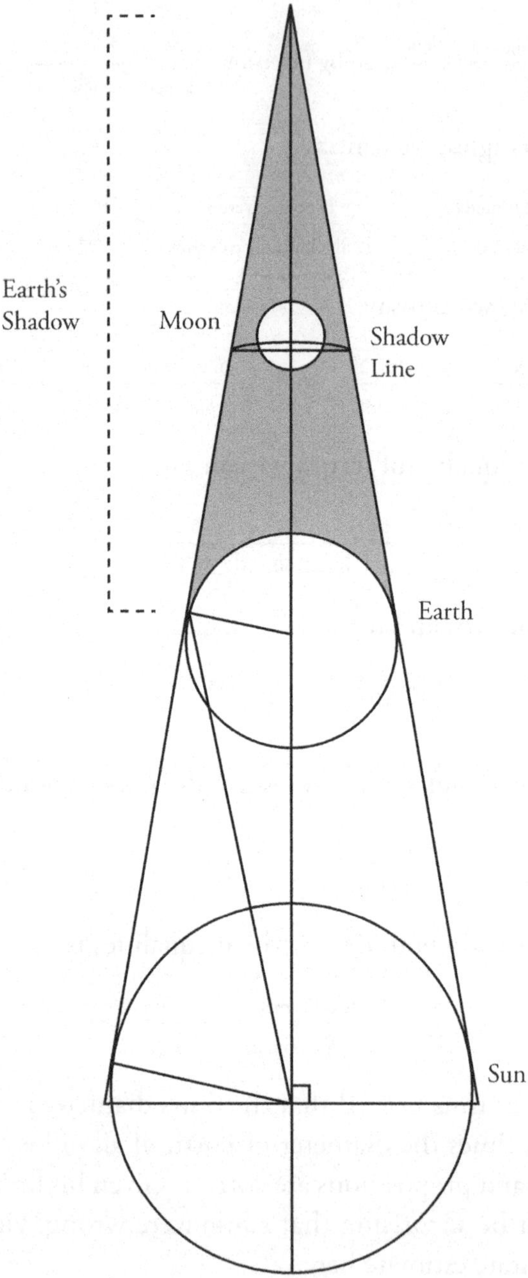

Fig. 1. Earth-Moon-Sun in Lunar Eclipse

The Russian Peasant Method

The earliest printed reference to the Russian Peasant Method of multiplication appears to be a reader's submission to a French mathematics journal in 1896.[1] Gason Albert Gohierre de Longchamps (1842–1906), the journal's editor, queries the readership for a proof of the method described by one Mr. Plackowo, a resident of the Russian province of Tambov. The French edition (1907) of an English popular mathematics book cites this French journal as the source for its own discussion of the Method.[2] In 1910, another French periodical published a demonstration, claiming—in turn—this English popular mathematics book as its source.[3]

The first English-language reference to the Method appears in 1910: Joseph Bowden, Professor of Mathematics at Adelphi College in Brooklyn, presented a paper on the subject to the American Mathematical Society's meeting in New York City.[4] He

---

1 *Journal de Mathématiques Élémentaires,* Vol. 5, Ser. 4 (Librairie Ch. Delagrave, 1896), pp. 22-23, 37. For a discussion of this journal's primacy in the order of publication, see Raymond Clare Archibald, "Undergraduate Mathematics Clubs," *The American Mathematical Monthly,* Vol. 25, No. 3 (March 1918), pp. 132-142; p. 141.

2 W. W. Rouse Ball, *Récréations Mathématiques et Problèmes des Temps Anciens et Modernes* (Librairie Scientifique A. Hermann, 1907), pp. 52-53. Curiously, the English editions of this book do not include the Russian Peasant Method. The demonstration of the Method appears in the first chapter only in the French edition, a unique contribution authored by J. Fitz-Patrick, the book's translator.

3 See the anonymous note, "La multiplication russe," *Cosmos* (November 12, 1910), p. 560.

4 The AMS meeting was held on February 25, 1910. See Joseph Bowden, "The Russian peasant method of multiplication," *Bulletin of the American*

published a version of this paper in 1912, crediting the Method's multilingual transmission to David Eugene Smith, the intellectual godfather of the Science Venerable Press.[5] Bowden defines the Method as follows:

> Having given the positive integers **a** and ***b***, to multiply **a** by ***b*** write down **a x *b***; under **a** write the exact or lower quotient obtained by dividing it by **2**; under this quotient write the quotient obtained by dividing it by **2**, and so on, until you obtain the quotient **1**.
>
> Under ***b*** write its double, under this double its double, and so on, until you have as many numbers in the second column as in the first.
>
> Next add the numbers in the second column which correspond to odd numbers in the first.
>
> The result is the product of **a** and ***b***...

For example, to multiply 13 by 24, successively halve the first term, discard the remainders, and double the second term in parallel columns:

Mathematical Society, Vol. XVII, No. 8 (May 1911), pp. 395-396. Bowden's paper was also summarized as a note in *Nature* (June 27, 1912), p. 431.

5 Joseph Bowden, "The Russian Peasant Method of Multiplication," *The Mathematics Teacher,* Vol. 5, No. 1 (September 1912), pp. 4-8. For an extremely brief summary of the Russian Peasant Method, see David Eugene Smith, *History of Mathematics: Special Topics of Elementary Mathematics, Vol. II* (Ginn and Company, 1925), p. 106.

1st Column	Action	2nd Column	Action
13 (odd)	Divide by 2 = 6.5, Discard 0.5 = 6	24	Multiply by 2 = 48
6	Divide by 2 = 3 No remainder to discard	48	Multiply by 2 = 96
3 (odd)	Divide by 2 = 1.5 Discard 0.5	96	Multiply by 2 = 192
1 (odd)		192	

From the second column, add all the terms that align with an odd number in the first column:

$$24 + 96 + 192 = 312.$$

Newton's Moon Test Error

Forsyth recounts the story of how Sir Isaac Newton initially computed just under 14 feet as the distance the Moon "falls toward" the Earth in 1 minute of orbit. If his theory of gravitation was correct, however, the value should be just over 16 feet. Newton apparently, simply, and mistakenly assumed that one degree of latitude measured 60 miles instead of 69.5 miles.[6] But how does this measurement affect the final result? As with many things related to Isaac Newton, what *exactly* happened here is a bit murky.[7] We can, nevertheless, reverse-engineer some of the mathematics.[8]

---

6 In Proposition XIX of Book III of the *Principia*, Newton writes: "Our countryman, Mr. Norwood, measuring a distance of 905,751 feet of London measure between London and York, in 1635, and observing the difference of latitude to be 2° 28', determined the measure of one degree to be 367,196 feet of London measure...." If one mile is 5,280 feet, then there are 69.5 miles in one degree of latitude, by Norwood's measurements.

7 David Brewster, *Memoirs of the Life, Writings, and Discoveries of Sir Isaac Newton* (Thomas Constable and Co., 1855), p. 26: "Knowing the force of gravity at the earth's surface, he was, therefore, led to compare it with the force exhibited in the actual motion of the moon, in a circular orbit; but having assumed that the distance of the moon from the earth was equal to sixty of the earth's semidiameters, he found that the force by which the moon was drawn from its rectilineal path in a second of time was only 13.9 feet, whereas at the surface of the earth it was 16.1 in a second."

8 The reverse-engineering is due entirely to the following sources: John Cox, *Mechanics* (Cambridge University Press, 1909), pp. 92-93; W. W. Rouse Ball, *An Essay on Newton's Principia* (Macmillan and Co., 1893), pp. 13-17; William L. Harper, *Isaac Newton's Scientific Method: Turning Data into Evidence about Gravity & Cosmology* (Oxford University Press, 2011); and Douglas W. MacDougal, *Newton's Gravity: An Introductory Guide to the Mechanics of the Universe* (Springer, 2012). This "moon test" is discussed in Newton, *Principia*, Book III, Proposition IV, Problem IV, and Proposition XXXVII, Problem XVIII, Corollary 7.

In the accompanying chart (Fig. 2), which is not drawn anywhere close to scale, we have the circular orbit of the **Moon** with the **Earth** at its center. In our scenario, the Moon travels along the ***Arc*** formed by angle θ, from **Moon Start** to **Moon End**, in 1 minute. We can take several things as given:

- The radius of the **Earth** is given as ***R***. This value is a variable, for now.
- The radius of the orbit of the **Moon** is ***60R***.
- By definition, the diameter of the orbit of the **Moon** is ***60R x 2 = 120R***.
- By definition, the circumference of the orbit of the **Moon** is ***120Rπ***.
- By definition, there are 360 degrees in a circle.
- By definition, the distance from point ***A*** to **Earth** is ***cos(θ)***.
- By observation, there are 39,343 minutes in a lunar period.[9]
- By definition, there are 5,280 feet in a mile.

Next, we can determine the length of the ***Arc*** formed by angle θ:

$$\textit{Length of Arc} = \frac{\textit{circum. of Moon's orbit}}{\textit{1 lunar period}} = \frac{120R\pi}{39{,}343 \textit{ minutes}}$$

Finally, we can make two approximations, which become more accurate the smaller θ becomes:

- The diameter of the Moon's orbit, ***120R***, is nearly equal to the distance from point **A** to point **B**. For our purposes, they are equal.
- The distance from **Moon End** to point **A** is nearly

9 In *Principia*, Newton assumes 27 days, 7 hours, 43 minutes. This is the sidereal period rather than the synodic period, which is from one full moon to the next.

equal to the length of the ***Arc***. For our purposes, they are equal.

Our goal is to determine the value of $\boldsymbol{X}$ in terms of $\boldsymbol{R}$. This measures the distance that the **Moon** "falls toward" **Earth** during 1 minute of its orbit along the ***Arc***:

$$\boldsymbol{X} = \frac{\textit{(distance Moon End to point A)}^2}{\textit{distance point A to B}} = \frac{(\text{length of Arc})^2}{\text{diameter of Moon's orbit}}$$

$$= \frac{(120R)\pi^2}{(39{,}343\ mins)^2} = \frac{(120R)\pi^2}{(39{,}343\ mins)^2}$$

But what is ***R***, the Earth's radius? If Newton assumed that one degree of latitude was 60 miles long on the Earth, then, given 360 degrees, this would imply a circumference of 21,600 miles. The circumference is equal to twice the radius multiplied by π and that, therefore, the radius is equal to 10,800 miles divided by π, or 3,437.7467 miles. Given 5,280 feet in a mile, R is equal to 18,151,302.9497 feet. Substituting this value for ***R***:

$$\boldsymbol{X} = 13.8889\ \textit{feet}$$

Alternatively, we can use this value of ***R*** to derive the radius of the Moon's orbit, and combine it with a semi-obscure trigonometric function called the versed sine, which was familiar to Newton.10 We don't compute the versed sine of angle ***θ*** directly. Instead:

$$\textit{versin}\ (\boldsymbol{\theta}) = \mathbf{1} - \mathbf{cos}\ (\boldsymbol{\theta})$$

---

10 For a brief history of the versed sine, see David Eugene Smith, *History of Mathematics: Special Topics of Elementary Mathematics*, Vol. II (Ginn and Company, 1925), pp. 618–619.

We can now say:

$$X = (\textit{radius of the Moon's orbit}) \times \textit{versin}\ (\boldsymbol{\theta}) = \mathbf{60}\boldsymbol{R} \times [\mathbf{1} - \mathbf{cos}\ (\boldsymbol{\theta})]$$

Assume:

$$\boldsymbol{\theta} = \frac{\textbf{\textit{360 degrees}}}{\textbf{\textit{39,343 minutes}}} = \textbf{\textit{0.00915 degrees per minute}}$$

Given that our scenario involves 1 minute of the Moon's orbit, the radius of the Moon's orbit is ***60R***, and ***R*** equals 18,151,302.9497 feet, then:

$$X = (1{,}089{,}078{,}176.98\ \textit{feet}) \times [1 - \textit{cos}\ (\mathbf{00.915\ degrees})] = 13.8889\ \textit{feet}$$

Either method of calculation can be reworked using 69.5 miles as the value of ***R***, yielding:

$$\boldsymbol{X} = 16.08746\ \textit{feet}$$

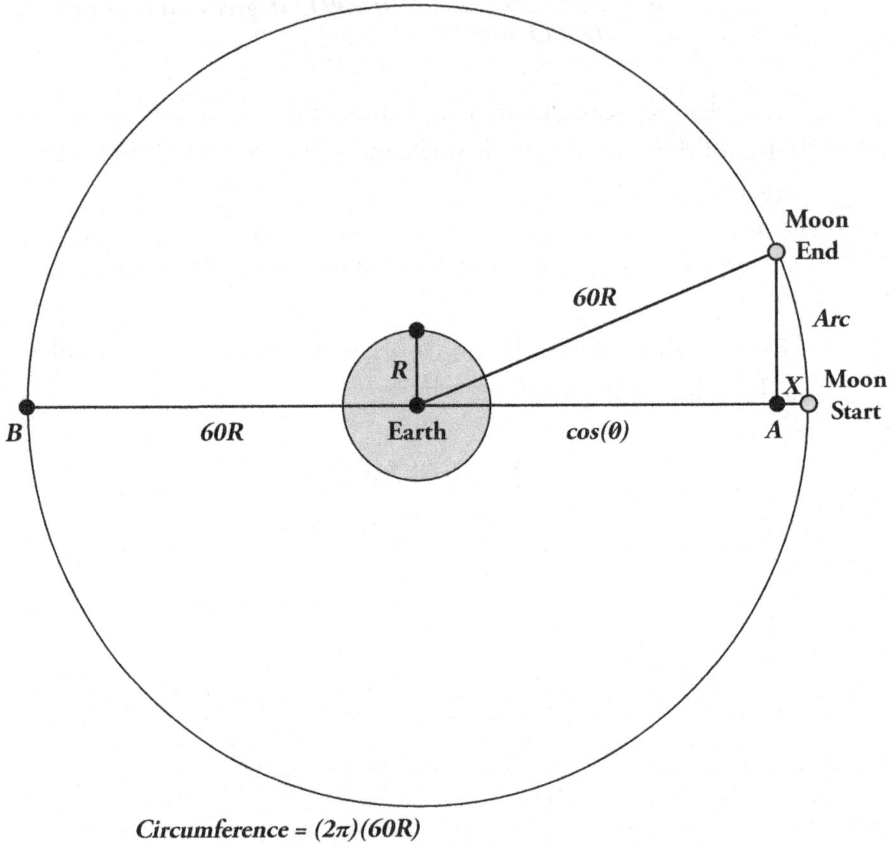

Fig. 2. Newton's Moon Test

Archibald Read Richardson

Archibald Read Richardson (1881–1954) was a veteran of the Second Boer War and the First World War. He served as a battalion commander and was wounded in 1918 during Operation Michael, Germany's final major offensive. D.S.O. refers to Distinguished Service Order, an award for gallantry. The Royal Society's biographical sketch reads:

> Few men could have survived the severe wounds and consequent operations which he suffered. . . . It was then that his true quality was displayed, in the singleness of purpose with which he turned his immense vitality to the channels of teaching and research.[11]

Richardson and Forsyth overlapped on the Imperial College faculty in the 1912–1914 period:

> Richardson, who had known Forsyth since Imperial College days, frequently spoke of what he owed to his teacher: the respect was mutual and Forsyth made frequent visits to Swansea for the purpose of meeting and staying with Richardson. Perhaps it was Forsyth who first aroused in Richardson the sense of propriety in a mathematical investigation—of not merely gaining the result, but gaining it by the right treatment.[12]

---

11 H. W. Turnbull, "Archibald Read Richardson, 1881–1954," *Biographical Memoirs,* Vol. 1 (Royal Society, November 1951), pp. 222–237; p. 224.
12 Ibid., p. 227.

Frederick Donnan Controversy

In September 1928, Frederick George Donnan (1879–1956) delivered an evening address to the British Association for the Advancement of Science in Glasgow. In the address, he cites the recent work of Archibald Vivian Hill (1886–1977), who won the Nobel Prize in Physiology in 1922:

> Now it is just here, at the very gate between life and death, that the English physiologist, A. V. Hill, is on the eve of a discovery of astounding importance, if indeed he has not already made it. . . . The living cell is like a battery which is constantly running down and which requires constant oxidation to keep it charged. It is perhaps a little premature at the present moment to say how far these results will prove to be general. Personally, I believe that they are of great importance and generality, and that for the first time in the history of science we begin, perhaps as yet a little dimly, to understand the difference between life and death, and therefore the very meaning of life itself.[13]

For his part, Hill "laughed incredulously" when told of Donnan's address:

> Donnan is an old friend of mine, and some time ago I lent him a paper containing the results of my researches. He has read into what I wrote far more than I, at any rate, would read. I certainly don't think that anything I've done means that we shall shortly

13 Donnan, "The Mystery of Life," *Report of the Ninety-Sixth Meeting* (British Association, 1929), pp. 659–666.

> be able to plumb the secret of life. Life itself, the fundamental thing, will, I think, always remain secret. It has been said that I am able to make a living cell in a laboratory. That is nonsense.[14]

At the time of the speech, Donnan does not appear to have been a sectional president. He chaired Section B (Chemistry) of the British Association's meeting in Liverpool in 1923.[15]

---

14 See "England Stirred by Theories on Life," *New York Times* (September 13, 1928).

15 See Francis Arthur Freeth, "Frederick George Donnan, 1879–1956," *Biographical Memoirs,* Vol. 3 (Royal Society, November 1957), pp. 23–39, and Bernard Katz, "Archibald Vivian Hill," *Biographical Memoirs,* Vol. 24 (Royal Society, November 1978) pp. 70–149.

The Science Venerable

Voltaire once remarked—"One merit of poetry few will deny; it says more and in fewer words than prose." With equal significance we may say, "One merit of mathematics few will deny; it says more and in fewer words than any other science." The formula $e^{\pi i} = -1$ expresses a world of thought, of truth, of poetry and of religious spirit, for "God eternally geometrizes."

— Creed of David Eugene Smith

Made in the USA
Monee, IL
05 November 2026

67375672R00049